The World of Steel

Joachim Schlegel

The World of Steel

On the History, Production and Use of a
Basic Material

Joachim Schlegel
Hartmannsdorf, Germany

ISBN 978-3-658-39732-6   ISBN 978-3-658-39733-3   (eBook)
https://doi.org/10.1007/978-3-658-39733-3

Responsible Editor: Frieder Kumm
This Springer imprint is published by the registered company Springer Fachmedien Wiesbaden GmbH, part of Springer Nature.
The registered company address is: Abraham-Lincoln-Str. 46, 65189 Wiesbaden, Germany

The most beautiful thing we can experience is the mysterious.

Albert Einstein

Foreword

The material steel, with its versatile properties and an almost incalculable wealth of applications, accompanies us all day long – and has done so for many generations.

As we read this, we may be stirring our coffee with a stainless steel spoon looking at the number combination 18/10 on its back. The exact designation is X5CrNi18-10. During manufacturing, it is listed the number 1.4301. In the workshop, it is named V2A. Directly, we find overselves in the middle of the world of steel, where "The World of Steel" accompanies us with an unmistakable and pleasant writing style. It shows that teaching about steel is not difficult to understand and is definitvely not as far away from us as we might think.

A great strength of this book is the successful mixture of technically well comprehensible contents combined with a huge amount of versatile, practical applications. Therefore, this book is suitable for many users: For basic teaching in the discipline of materials science and production technology as well as for technical postgraduate studies based on a business degree.

In addition to teaching the basics, author Dr Joachim Schlegel takes us back through the history of steel, covers the area up to the latest applications, and amazes us with the question "Did you know ...?".

The book "The World of Steel" will take its place linking theory and practice, especially for those who enjoy interesting, visual and practice-oriented learning and teaching.

I wish the author and his work all the best.

Prof. Dr.-Ing. Andreas Zilly
Professor of Materials Science at the Duale
University of Baden-Württemberg

Preface

A stainless steel sink in the modern kitchen is ubiquitous; so is the stainless steel cutlery and the often used steel wool to clean pots. On steel skates, the bobsleigh glides towards the target on the ice channel. It is fun to play boules, translated from French "balls", which today are made of steel. Needles for knitting, weaving, felting, combing and sewing, whether for the housewife or for use in textile machines, are made of special steels. And if you want to build something in your own home, you will be pleased to have suitable tools made of durable tool steel, such as a spiral drill, a jigsaw, a chisel or hammer. Steel stamps emboss coins or form medicine into pills. A copper-plated steel wire is used in fireworks. Bowden cables on bicycles for operating the brakes up to suspension cables for suspension railways or suspension bridges are made of high-strength steel wire. Even if you can't see it, there is a lot of steel in modern concrete buildings. Steel rails for railways and trams secure our mobility. The power lines are attached to steel lattice masts. Steel beams carry loads, and modern cars consist of many different types of steel. No matter where we are and what we are doing, steel is always part of our lives, at work or in our leisure time, sometimes as a work of art. Steel is essential, recyclable and has a very special meaning: In our modern industrial society, steel is the basic material for all important industrial sectors, such as automotive and shipbuilding, aerospace industry, apparatus and machine construction, bridge and steel construction, energy and environmental technology, packaging industry, household and sports industry, medical technology, robotics and IT technology, etc. All the global megatrends of today, such as energy supply, mobility, healthcare, environmental and climate protection, cannot be solved and mastered without steel.

The large-scale production and processing of steel is state of the art today and the number of developed steels is impressive: Already over 2500 steel grades are listed in the European steel register. The over 5000 year long history of iron and steel production is rather impressive as well. The world of steel is constantly evolving and has become so diverse and complex in the meantime that it is not easy to overview it in practice. The reader should be brought closer to this world, from steel production, further processing with ingot and continuous casting, forming and machining to finishing, testing and packaging of the products, the processes and facilities used for this purpose, including the environmentally friendly recycling and disposal of waste. Interesting and new techniques and applications will also be pointed out.

Not high-scientific and all-encompassing, but informative and exciting, structured, above all understandable and with concrete practical examples, partly also with historical references - this is how an insight into the world of steel will be granted here. A timeline provides information on important milestones in iron and steel production in chronological alignment with social and technical events. And in a glossary, finally, terms and abbreviations from the practice of steel metallurgy, steel processing and material testing are explained in order to find one's way around the world of steel more quickly.

I would like to thank the shareholders of the BGH Edelstahl Group, in particular Messrs. Rüdiger and Sönke Winterhager, who promoted the creation of this book. The BGH Edelstahlwerke GmbH kindly provided some photos from production and approved the publication of photos that I took during my work in the companies of the BGH Group for training purposes.

I would like to thank Prof. Dr.-Ing. Andreas Zilly, Professor of Materials Science at the Duale University of Baden-Württemberg, Stuttgart, for his expert support in the preparation and review of the manuscript. Valuable hints on steel production and the chronology of iron production and processing were given by Dr.-Ing. habil. Bernd Lychatz, Institute of Iron and Steel Technology at the TU Bergakademie Freiberg. I would like to thank Mr. Frieder Kumm M.A., Senior Editor, Department of Civil Engineering of Springer Vieweg, for his motivation and support during the creation of the manuscript and the design of the book. Finally, Mr. Claus-Dieter Bachem, Project Coordinator at Springer Nature, and Mr. Georg Haller-Kaimann, Implementation Manager at Springer Nature, have contributed greatly to translating this with book into English with help of an automatic translation software using artificial intelligence.

I would like to thank my brother, Dr.-Ing. Christian Schlegel, and my son, Dr. Peter Schlegel, for their help with proofreading the manuscript in German and in English, respectively. And I would also like to sincerely thank my dear wife Birgit for always having my back and also for her critical remarks on understandable formulations.

Dr.-Ing. Joachim Schlegel

The original version of the book was revised: For detailed information please see Correction. The correction to the book is available at https://doi.org/10.1007/978-3-658-39733-3_14

Contents

Systematization of Materials

▶ The history of mankind is closely linked to the development and use of materials, which even shaped different periods in history such as the Stone Age, Bronze Age and Iron Age. It was a long, hard way: from the stone handaxe to the use of processed, solidified metals such as gold, silver and copper, the discovery of the first alloy (bronze) to today's targeted material development. Metallurgical processes and plants, processing technologies and usage concepts had to be developed and implemented. The driving force for this was the constantly increasing demand for materials associated with the progress of mankind.

"Materials made to measure" – durable, sustainable, light, but highly resilient, recyclable, and even intelligent – such materials can already be produced today. Yet the requirements continue to increase. The complex properties of a material must be considered for its future use, as shown for example in Fig. 1.1 (Weißbach et al., 2018), e.g.:"

- *State of the material*
- *Interactions with other materials*
- *Behavior under mechanical stress*
- *Behavior during manufacturing (forming, machining, coating, etc.)*
- *Behavior under environmental influences*

These complex requirements mean that materials have been mainly divided according into chemical groups, since these essentially also determine the characteristic properties of the material concerned (Briehl, 2014; Kutz, 2013). Today, materials are divided into the following main groups: metallic systems, semiconductors and non-metallic systems (Fig. 1.2).

J. Schlegel, *The World of Steel*, https://doi.org/10.1007/978-3-658-39733-3_1

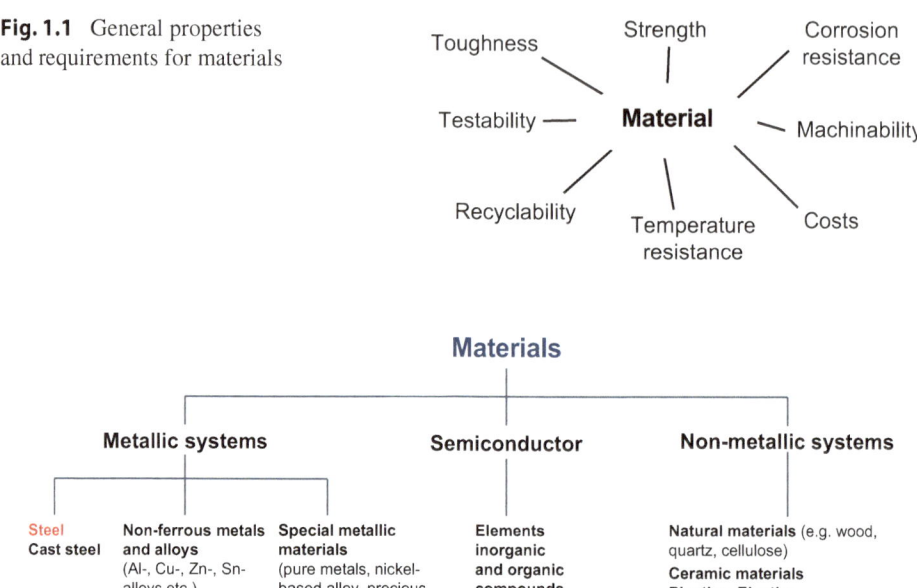

Fig. 1.1 General properties and requirements for materials

Fig. 1.2 Division of materials into main groups

The most important material groups are characterized below according to their properties.

Metals (75% of all chemical elements).
Metals have a metallic bond, are usually shiny, plastic, malleable, alloyable, meltable, ductile, hard or tough, weldable and recyclable.
Main materials:
Steels, cast iron, aluminum, copper, nickel and titanium alloys, pure metals, precious metals, special alloys, nickel-based alloys.
Examples:
Structural steel and metalwork, lightweight construction, toolmaking, machine and plant construction, vehicle construction, shipbuilding, medical engineering, textile machinery construction, electrical engineering, electronics, hydraulics / pneumatics, and many others.

Metals are to be found in all areas of our lives. For example Fig. 1.3 shows an interesting steel structure of the Sony Center at Potsdamer Platz in Berlin with fabric and 105 t of safety glass.

Semiconductor
These are materials of electrical engineering, electronics and information technology, which are conductive between the metallic conductors and the insulating ceramics or polymer materials. They are functional materials with a huge social impact on humanity.

Fig. 1.3 Steel roofing struture of the Sony Center at Potsdamer Platz, Berlin. (Photo: Schlegel, J.)

Main materials:
Silicon, germanium, gallium arsenide.
Examples:
Components for electrical engineering and electronics, optoelectronics, LED lighting technology, mobile radio technology, etc., produced using semiconductor materials. Figure 1.4 shows a monitor module with components made of semiconductor materials.

Plastics
Their properties depend on the manufacturing process, the additives and the temperature conditions: insulating, light, easily moldable, chemically resistant, low temperature resistant, colorable, transparent to opaque, rubbery to brittle, high thermal expansion, difficult to recycle, low strength and usually inexpensive.
Main materials:
Polyvinyl chloride, polyethylene, polypropylene, polystyrene, polyamide, polyethylene terephthalate.
Examples:
Plastic is omnipresent in our daily lives, and is a material for countless, indeed almost all, industrial sectors, for the areas of household, leisure, sports, etc. It is therefore superfluous to list all its applications; a fine example is shown in Fig. 1.5 in the form of a child's toy.

Fig. 1.4 View of a monitor module. (Photo: Schlegel, St.)

Glass and Ceramics

Glass and ceramics are usually mentioned together because they have similar properties: insulating, heavy, hard, wear-resistant, brittle, chemically resistant and temperature-resistant, transparent to opaque and non-toxic.

Main materials:

Clay minerals, silicides, oxides, carbides, nitrides, borides, hard materials

Examples of glass applications:

Glazing on buildings (windows, doors, facades, canopies, other glazing), automotive industry, household goods, beverage industry, works of art, etc. An example of a one-time and especially creative use of glass is the Hundertwasser toilet in Kawakawa on New Zealand's North Island. It has a glass-bottle wall, shown in Fig. 1.6. In the masonry, the famous artist *Friedensreich Hundertwasser* (1928–2000) set colorful glass bottles, a fanciful peculiarity that he cultivated in his architecture.

Examples of ceramic applications:

Cutting materials, parts and linings for plants and apparatus in the chemical industry, insulators, wearing parts, coatings and many other parts in electrical engineering/electronics. Ceramic ware is to be found in almost every household, e.g. in the form of dishes, vases and planters (Fig. 1.7).

Fig. 1.5 Fire truck turntable made of plastic. (Photo: Schlegel, J.)

Fig. 1.6 Detail of the glass-bottle wall of the Hundertwasser toilet in Kawakawa, New Zealand, built in 1999. (Photo: Schlegel, J.)

Fig. 1.7 Typical ceramic
planters (Photo: Schlegel, J.)

Composite Materials

Custom-made structural materials with unusual, multifunctional property combinations
are considered to be composite materials. They have high strength and stiffness, are light
and durable, but usually expensive, difficult to repair and not recyclable.

Types of composite materials:

- Fiber-reinforced composites
- Particle-reinforced composites
- Layer (sandwich) composites

Figure 1.8 shows these three types of composite materials schematically.
   The most important composite materials known today are:

- fiber-reinforced plastics
- carbon fiber-reinforced plastics
- metal matrix composites
- new material combinations

Types of composite materials

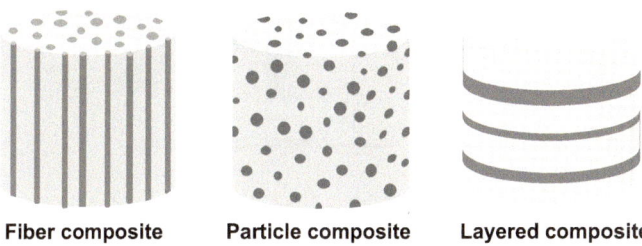

| Fiber composite | Particle composite | Layered composite |

Fig. 1.8 Schematic representation of the structure of composite materials

Examples:

Parts (mostly lightweight) for the automotive, aerospace and shipbuilding industry, space technology, motorsport, model making, sports equipment, medical technology, and many more.

If you also add natural materials, such as leather, you quickly come to a product that combines all of the above-mentioned material groups into a form of outstanding engineering, which is beautiful and very mobile, and more or less sustainable: our automobile ("self-propelled", motor vehicle).

The materials mentioned, metals, plastics, glass and ceramics, have two different solid states, see Fig. 1.9. These depend on the conditions during solidification, i.e. on the temperature gradient during the transition from the liquid to the solid state (Hornbogen et al., 2019).

Fig. 1.9 Representation of the material states crystalline and amorphous

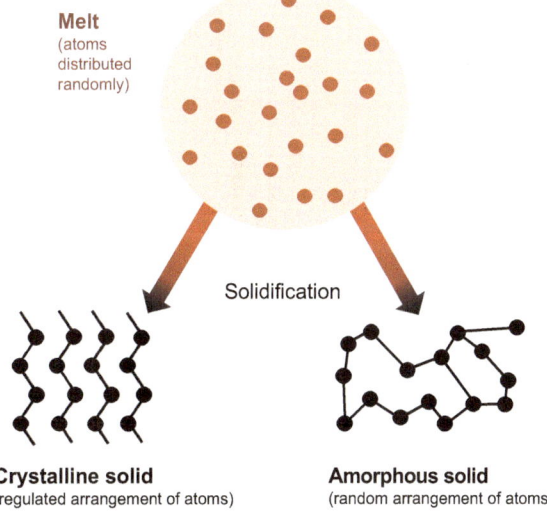

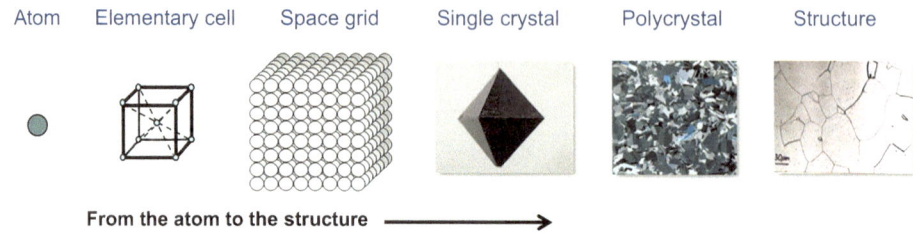

Fig. 1.10 From the atom, the smallest building block, to the polycrystalline structure of a steel

The crystalline state Metals, crystalline ceramics, semi-crystalline plastics.
In the crystalline state there is a regular arrangement of the building blocks (atoms), and it is a stable state. Solidification takes place abruptly with a phase transition at a defined melting point (temperature of the liquid – solid transformation).

The amorphous state glass, plastics, metallic glass
An amorphous state is to be understood as a "supercooled melt" (e.g. as in glass). The atoms are in an irregular arrangement. The density of this state is lower than in the crystalline state.

Technical Metals are almost exclusively *polycrystalline* (Eisenkolb, 1958 ff.). They form in the solid state crystals, which in turn consist of atoms in different spatial arrangements (lattice). The crystals are also called grains and their association structure. An interesting look inside a metal (Fig. 1.10): From the structure of a usable object, such as a drill, a crankshaft or a connecting rod, down to the smallest building blocks, namely the atoms.

Note

The following explanations only apply to the material *steel*; namely steel as the "forgeable", malleable iron with less than 2.06% carbon content (i.e. not cast iron).

References

Briehl, H. (2014). *Chemie der Werkstoffe*. Springer Vieweg.
Eisenkolb, F. (1958). *Einführung in die Werkstoffkunde. Bd. 1 – Allgemeine Metallkunde, Bd. II – mechanische Prüfung metallischer Werkstoffe, Bd. III – Eisenwerkstoffe*. Verlag Technik.
Hornbogen, E., Warlimont H., & Skrotzki, B. (2019). *Metalle: Struktur und Eigenschaften der Metalle und Legierungen*. Springer Vieweg.
Kutz, K.-H. (2013). *Struktur und Eigenschaften – Werkstoffe nach Maß*. Universität Rostock. Zentrum für Qualitätssicherung in Studium und Weiterbildung.
Weißbach, W., & Dahms, M. et al. (2018). *Werkstoffe und ihre Anwendungen: Metalle, Kunststoffe und mehr*. Springer Vieweg.

Steels always fascinate with their many, often also extraordinary properties, with a great potential for application and increasing production worldwide.

- What characterizes the material steel?
- How are steels classified or differentiated?
- What are steels composed of?
- What properties do steels have and how can they be influenced?
- Where are steels mainly used?

The following section provides an orientation with explanations of the basics of the material steel, the possibilities of classifying steels as well as short portraits of selected steel grades.

2.1 What is Steel?

Steel is malleable, forgeable iron with a carbon content of less than 2.06 mass-% (usually with <1 mass-% carbon). At carbon contents above 2.06 mass-%, one speaks of cast iron. Therefore steel is an iron-carbon compound. This is mixed with other metallic and non-metallic elements (alloyed) to obtain steels with different properties (Berns & Helmreich, 1980).

  Density of steel: approx. 7.85 to 7.87 g/cm^3.

  Melting point (iron/steel): depending on the chemical composition up to 1536 °C.

  The material steel is polycrystalline, that is, it is made up of individual crystal lattices. Their modifications are determined by the *base element iron*. Iron as the main

---

The original version of the chapter has been revised. A correction to this chapter can be found at https://doi.org/10.1007/978-3-658-39733-3_14

component of steel occurs in two types, namely as a space-centered cubic and a face-centered cubic crystal lattice (Bleck, 2010).

The space-centered cubic lattice (α-iron)

In this type of lattice, there is one iron atom at each of the eight vertices, and a ninth iron atom exactly in the center of the cube (Fig. 2.1).

In such an α-iron, which exists at temperatures up to 911 °C, a maximum of 0.018 mass-% of carbon can be dissolved. This resulting crystal consisting of iron and carbon atoms, that is, of different atoms, is called a mixed crystal. Since the iron in this case is in the form of a body-centered cubic lattice (α-iron), this mixed crystal is called α-mixed crystal and is also referred to as a *ferrite* (lat. "Ferrum", the "iron").

The face-centred cubic lattice (γ-iron)

There are atoms at the eight corners of the cube lattice. However, the center of the cube remains free. More atoms are arranged in the center of each of the six cube faces, as shown in Fig. 2.2.

In this γ-iron, which only occurs in the temperature range from 911 to 1398 °C, a maximum of 2.1 mass% carbon is dissolved. This γ-mixed crystal is called *austenite*,

Fig. 2.1 The space-centered cubic lattice

Space-centred cubic lattice (α-Iron)

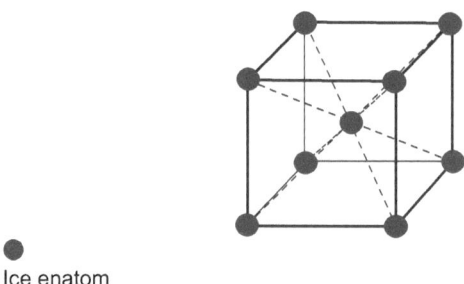

Ice enatom

Fig. 2.2 The cubic-centered cubic lattice (cubic close-packed)

Face-centered cubic lattic (γ-Iron)

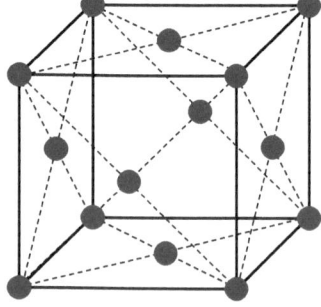

Iron atom

after Sir *William Chandler Roberts-Austen* (1843–1902). He was a British metallurgist who studied the physical properties of metals.

In iron-carbon alloys, the carbon is stored on interstitial sites in the iron crystal lattice. This results in so-called mixed crystals. These contain atoms of different alloying elements in the lattice structure. Depending on the crystal structure, the carbon is only slightly soluble in the iron lattices. When heated and cooled slowly, the iron atoms migrate, and a lattice transformation (phase transformation α into γ or γ into α) takes place. This process is deliberately used during heat treatment to create certain microstructures and thus corresponding properties of the steels.

▶   **Note** There is a variety of steels that have a lattice transformation, but also
      steels such as ferritic and austenitic steels that do not show such a transforma-
      tion.

2.2   The Iron-Carbon Diagram

The most important alloying element in steel is carbon (C). This is present as a compound (cementite – Fe_3C). In general, the higher the carbon content, the harder the steel, but also the more brittle. With the well-known iron-carbon diagram, as shown in Fig. 2.3, the phase composition of steel is described as a function of carbon content and temperature.

The mass percentages of carbon content are plotted on the x-axis below, and the temperature in °C is plotted on the y-axis upwards. For technical use, the pure two-substance system iron-carbon is only considered up to a maximum of 6.67 mass-% carbon (cementite – Fe_3C) and at normal pressure. It is assumed that the cooling takes place very slowly with complete transformation processes.

The iron-carbon diagram therefore only provides information on equilibrium conditions. In this way, the expected state of an unalloyed steel with a certain carbon content at a specific temperature can be determined. The microstructural changes taking place during temperature changes can also be predicted (Klemm, 1973). Therefore, the iron-carbon diagram is also the most important basis for the heat treatment of steel.

Explanations of the iron-carbon diagram

Line ACD: *Liquidus line*
Above this line, the alloy is liquid. Below this line, solidification begins. (Primary crystallization) of the melt.

Line AECF: *Solidus line*
Below this line, the alloy is completely solidified.

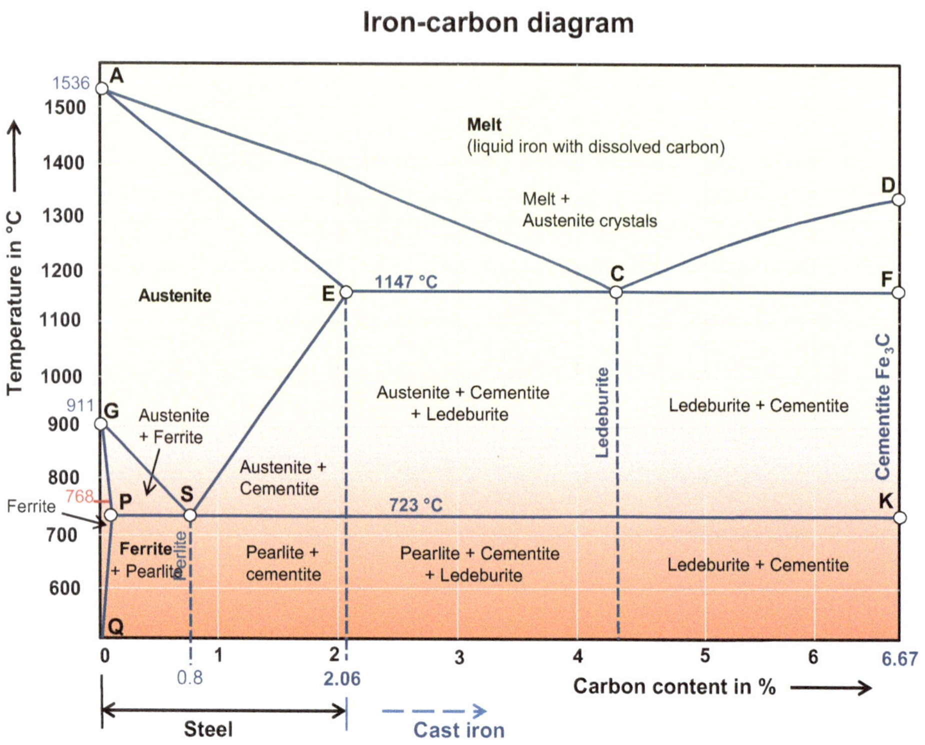

Iron-carbon diagram

Fig. 2.3 Simplified representation of the iron-carbon diagram

ECF line:

Above this line, liquid and solid phases exist side by side.

Line PSK:

This line describes the constant temperature of 723 °C. Below this line, the Austenite mixed crystals have completely disintegrated.

Line SE:

At carbon contents between 0.8 and 2.06 mass-%, the carbon in excess of solution as secondary cementite (Fe_3C) is precipitated during cooling.

GPQ line: It delimits the single-phase region of ferrite.

768 °C: *Curie temperature* (magnetic transformation) Up to this temperature, iron is ferromagnetic, above this temperature paramagnetic (non-magnetic).

Ferrite: α-mixed crystal (max. 0.018 mass-% carbon).

Austenite: γ-mixed crystal (max. 2.06 mass-% carbon).

Perlite and Ledeburite:

They are special phase mixtures (structures) that only occur during slow cooling. If cooling takes place quickly, e.g. by quenching in water, the austenite becomes a hard and brittle structure, *martensite.*

Cementite Fe$_3$C:

It is a microstructure phase with 6.67 mass-% carbon, which can occur in three different forms at the same composition:

They are special phase mixtures (microstructures) that only occur during slow cooling. If cooling is rapid, e.g. by quenching in water, a hard and brittle microstructure, called martensite, *Martensit* is formed from austenite.

Primary cementite: primary crystallization from the melt, line CD

Secondary cementite: precipitation from austenite, line ES

Tertiary cementite: precipitation from ferrite, line PQ

Example

The change in microstructure for pure iron and a 1 mass-% carbon steel during heating and cooling is shown in Figs. 2.4 and 2.5.

2.3   Designation System and Classification of Steels

▷   As numerous and diverse as the steels, so are the names: steel type, steel grade, steel quality, steel name, steel brand, material number, brand name. Finding your way around in steel practice is not quite easy. For the identification and assignment of steels, rules are laid down in DIN EN 10027-1 for the designation by means of short names as well as in DIN EN 10027-2 for a numbering system. This designation system is described in detail, for example, in the Steel Key Pocket Book (Wegst & Wegst, 2019). Based on this, Fig. 2.6 shows a simplified representation of the steel designation system with examples of different steel grades.

Steel short names

They give indications of the use, the mechanical and physical properties or the chemical composition of the steels. The steel short names consist of main and additional symbols, which can be letters (e.g. chemical symbols) or numbers (for the content of the alloying elements) respectively. These specifications differ for unalloyed, alloyed and highly alloyed steels as well as for high-speed steels (Langehenke, 2007).

Unalloyed steels (quality steels) are indicated by the letter C for carbon, followed by the carbon content. The number given for the carbon content is always multiplied by 100. I.e. in order to recognize the actual content, this number must be divided by 100.

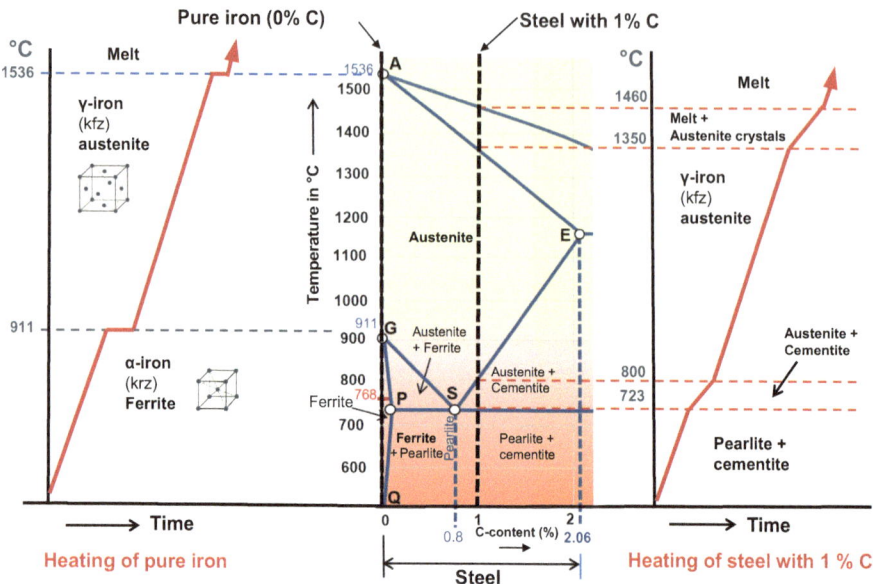

Fig. 2.4 Change in microstructure during heating of pure iron (0% carbon) and a 1 mass-% carbon steel

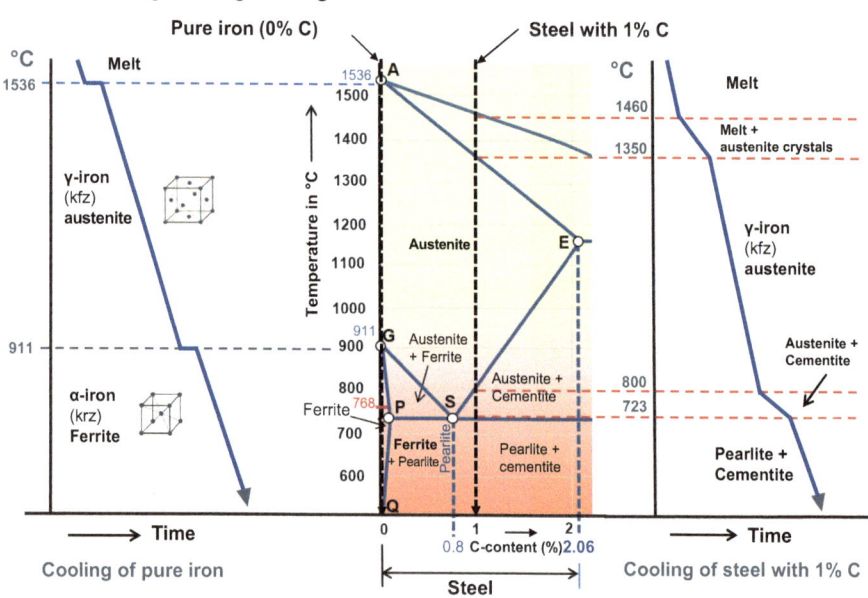

Fig. 2.5 Progression of microstructure changes during cooling of pure iron (0% carbon) and of a steel with 1 mass-% carbon

Steel grades	Main icons		Additional symbols
	Letter	C-content	
Unalloyed steels with manganese content ≤ 1 %			
C35E (1,1181)	**C** - Carbon	**35** (/100 = 0.35% C)	**E** - prescribed max. S content
Unalloyed steels with manganese content ≥ 1 %, unalloyed free-cutting steels, alloyed steels			
28Mn6 (1,1170)		**28** (/100 = 0.28% C)	**Mn** - Manganese **6** (/4 = **1,5% Mn**)
			Elements / Factor: Cr, Co, Mn, Ni, Si, Mn — 4; Al, Be, Cu, Mo, Nb, Pb, Ta, Ti, V, Zr — 10; Ce, N, P, S — 100; B — 1000
Alloyed, high-alloy steels			
X5CrNi18-10 (1,4301)	**X** - alloy steel	**5** (/100 = 0.05% C)	**CrNi18-10** (18% chromium, 10% nickel) *No factor!*
High-speed steels			
HS6-5-2 (1,3343)	**HS** - High Speed Steel [high-speed steel]	**6-5-2** (6% tungsten, 5% molybdenum, 2% vanadium) *Content of alloying elements in % in the order:* **W - Mo - V - Co**	

Fig. 2.6 Overview of the steel designation system

Example:
C15 – an unalloyed steel with 15/100 = 0.15 mass-% carbon.

In *unalloyed steels with manganese content ≥ 1 mass-%* the carbon content is given in first place, also multiplied by the factor 100 and in contrast to unalloyed steels with manganese content ≤ 1 mass-% always without the letter C. This is followed by the chemical symbols for the alloying elements and their mass content. It should be noted that these mass contents have always been multiplied by different factors. These multipliers for the individual alloying elements are as follows:

Factor 4: Chromium (Cr), Cobalt (Co), Manganese (Mn), Nickel (Ni), Silicon (Si), Tungsten (W)
Factor 10: Aluminium (Al), Beryllium (Be), Copper (Cu), Molybdenum (Mo), Niobium (Nb), Lead (Pb), Tantal (Ta), Titanium (Ti), Vanadium (V), Zirconium (Zr)
Factor 100: Cer (Ce), Nitrogen (N), Phosphorus (P), Sulfer (S), Carbon (C)
Factor 1000: Bor (B)

In order to recognize the actual alloy contents, the numbers given in the steel short name must be divided by the corresponding multipliers.
Example:
28Mn6 – an alloyed steel with 28/100 = 0.28 mass-% carbon and 6/factor 4 = 1.5 mass-% manganese.

High-alloy steels always have a mass fraction of various alloying elements of a total of at least 5 mass-%. These steels are characterized by an X at the beginning of the short name. This is followed by the carbon content, again multiplied by the factor 100 in principle, and the other alloying elements with their chemical symbols. The indication of the alloying elements takes place in the order starting with the highest content. This is followed by the respective mass fractions belonging to the alloying elements. However, these are not multiplied by a factor (typical for high-alloy steels!).
Example:
X5CrNi18-10 – a high-alloy, austenitic, non-corrosive stainless steel with 0.05 mass-% carbon, approx. 18 mass-% chromium and 10 mass-% nickel. This steel is the first commercial, non-rusting steel, which is also known as V2A and corresponds to the material number 1.4301.

An exception are the *high-speed steels*. Here a special designation system applies. In first place is the designation HS, followed by the mass fractions of the alloying elements in the prescribed order tungsten—molybdenum—vanadium—cobalt. The mass fractions of the individual alloying elements are given here in whole, rounded numbers.
Example:
HS 6-5-2 C – a standard high-speed steel with 6 mass-% tungsten, 5 mass-% molybdenum and 2 mass-% vanadium (corresponding to the material number 1.3343).

Material numbers

They are issued by the European Steel Registry and consist of the material main group number (first number with point), the steel group numbers (second and third number) and the sequence numbers (fourth and fifth number). In addition, the steelmaking process and the treatment condition can be characterized by two appendix numbers XX. Figure 2.7 shows this numbering system using the example of the 1.5920 – 18CrNi8 tool steel.

The appendix numbers XX are structured as follows:

Position 6: Steelmaking process

0 – indeterminate or without meaning
1 – unkilled Thomas steel
2 – killed Thomas steel
3 – other melting type, unkilled
4 – other melting type, killed
5 – unkilled Siemens-Martin steel
6 – killed Siemens-Martin steel
7 – unkilled oxygen-blown steel
8 – killed Oxigen blast steel
9 – electric steel

Material numbers system

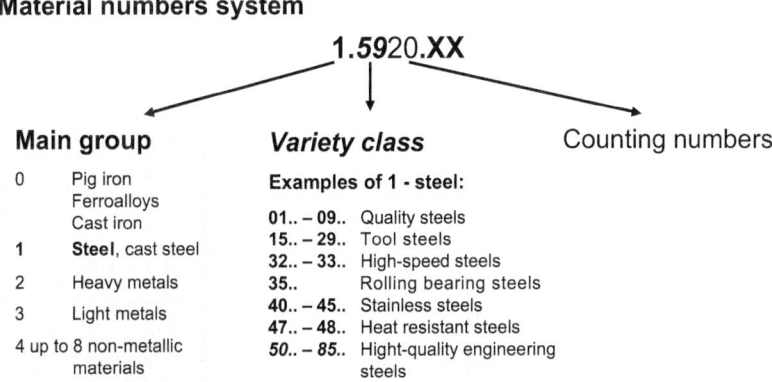

1.5920.XX

Main group   *Variety class*   Counting numbers

0	Pig iron
	Ferroalloys
	Cast iron
1	**Steel**, cast steel
2	Heavy metals
3	Light metals
4 up to 8	non-metallic materials

Examples of 1 - steel:

01.. – 09..	Quality steels
15.. – 29..	Tool steels
32.. – 33..	High-speed steels
35..	Rolling bearing steels
40.. – 45..	Stainless steels
47.. – 48..	Heat resistant steels
50.. – 85..	Hight-quality engineering steels

Fig. 2.7   Material numbering system using the example of the 1.5920—18CrNi8 tool steel

Position 7: Type of treatment

0 – no or any treatment

1 – normalized annealed

2 – soft annealed

3 – annealed for best machining

4 – tough hardened and tempered

5 – hardened and tempered

6 – hard annealed

7 – coldhardened

8 – cold hardened for springs

9 – treated according to special instructions

Brand Names

Some steels are still traded under the brand names legally protected by the steel manufacturers. Here are some current examples:

- Invar (iron-nickel alloy with 36 mass-% nickel, 1.3912, brand name of Aperam Alloys Imphy, France)
- Nirosta (stainless steel, brand name of Thyssen Krupp Nirosta)
- Cromargan (high-quality Cr-Ni steel, trade name of WMF)
- Inconel, Incoloy, Monel, Nimonic, Inco-Weld, Nilo, Brightray, Coronel, Udimet (protected brand names of Special Metals, USA)

In addition, there are historically grown special terms which are still in use in the steel world such as V2A (trial melt 2 austenite, Krupp patent), today the 1.4301 – X5CrNi18-10.

Comparison of international steel standards

Steel	USA AISI / SAE	Japan JIS	England BS	Germany DIN	Italy UNI	France AFNOR	Spain UNE
Carbon steel	1055	S55C	070M55	1.0511 C55	1C55	AF70C55	–
Engineering steel	4340 4337	SNcM447	817M40	1.6582 34CrNiMo6	35NiCrMo6KB	34CrNiMo8 35NCD6	F.1272
Tool steel	D2 H13	SKD11 SKD61	BD2 BH13	1.2379 X155CrVMoV12 1.2344 X40CrMoV5-1	X155CrVMo121KU X40CrMoV511KU Z40CDV5	X160CrMoV12-28 X40CrMoV5	F.520A F.5318
RSH (stainless, acid and heat resistant steel)	304 303	SUS304 SUS303	304S15 303S31	1.4301 X5CrNi18-10 1.4305 X10CrNiS18-9	X5CrNi1810 X10CrNiS1809	Z4CN19-10FF Z8CNF18-09	F.3504 F.3508

Fig. 2.8 Comparison of the most important international steel standards using some selected examples for carbon steel, structural steel, tool steel and for rust-, acid- and heatresistent steels

> **Note** Reference has to be made to international steel standardization: e.g. DIN EN (Europe), ASTM/AISI (USA), JIS (Japan). These are the operating fundamentals for the steel producing and processing industries and a consensus between industry, trade, research and consumers. For this purpose, Fig. 2.8 shows an overview comparing the most important international standards for selected steels (Marks & Tirler, 2016).

In practice, the *classification of steels* is usually carried out according to the following criteria:

- *Main quality classes*
- *Chemical composition*
- *Structure*
- *Application/properties*

Main quality classes of steels
Basic steels
These include all unalloyed steels for requirements for which no special measures are required in steel production.
Examples: Iron products, iron grills and railings, fences. Figure 2.9 shows, for example, a typical garden fence made of forged iron.

Quality steels
These consist of unalloyed and alloyed steel grades, which, for example, must meet certain requirements in terms of formability, weldability, deep drawability and grain size.